Managing Editor
Karen J. Goldfluss, M.S. Ed.

Editor-in-Chief
Sharon Coan, M.S. Ed.

Cover Artist
Barb Lorseyedi

Art Coordinator
Kevin Barnes

Art Director
Cjay Froshay

Imaging
James Edward Grace

Product Manager
Phil Garcia

Publisher
Mary D. Smith, M.S. Ed.

Practice Makes Perfect

Graphs & Patterns

GRADES 1 & 2

Authors

Teacher Created Resources Staff

Teacher Created Resources, Inc.

6421 Industry Way
Westminster, CA 92683
www.teachercreated.com
ISBN-13: 978-0-7439-3320-9
ISBN-10: 0-7439-3320-6
©2002 Teacher Created Resources, Inc.
Reprinted, 2006
Made in U.S.A.

Table of Contents

Introduction

The old adage "practice makes perfect" can really hold true for your child and his or her education. The more practice and exposure your child has with concepts being taught in school, the more success he or she is likely to find. For many parents, knowing how to help your children can be frustrating because the resources may not be readily available. As a parent it is also difficult to know where to focus your efforts so that the extra practice your child receives at home supports what he or she is learning in school.

This book has been designed to help parents and teachers reinforce basic skills with your children. *Practice Makes Perfect* reviews basic math skills for children in grades 1 and 2. The math focus is on graphs and patterns. While it would be impossible to include all concepts taught in grades 1 and 2 in this book, the following basic objectives are reinforced through practice exercises. These objectives support math standards established on a district, state, or national level. (Refer to the Table of Contents for the specific objectives of each practice page.)

- identifying what comes next in a series
- identifying shape/object and number patterns
- identifying missing shapes and numbers in a series
- classifying shapes
- reading and using pictographs to solve problems

- completing graphs
- reading and interpreting bar graphs
- interpreting information from a chart
- using charts, Venn diagrams, circle graphs, patterns, and tally sheets to solve problems

There are 36 practice pages organized sequentially, so children can build their knowledge from more basic skills to higher-level math skills. (**Note:** Have children show all work where computation is necessary to solve a problem. For multiple choice responses on practice pages, children can fill in the letter choice or circle the answer.) Following the practice pages are six test practices. These provide children with multiple-choice test items to help prepare them for standardized tests administered in schools. As your child completes each test, he or she can fill in the correct bubbles on the optional answer sheet provided on page 46. To correct the test pages and the practice pages in this book, use the answer key provided on pages 47 and 48.

How to Make the Most of This Book

Here are some useful ideas for optimizing the practice pages in this book:

- Set aside a specific place in your home to work on the practice pages. Keep it neat and tidy with materials on hand.
- Set up a certain time of day to work on the practice pages. This will establish consistency. An alternative is to look for times in your day or week that are less hectic and conducive to practicing skills.
- Keep all practice sessions with your child positive and constructive. If the mood becomes tense, or you and your child are frustrated, set the book aside and look for another time to practice with your child.
- Help with instructions if necessary. If your child is having difficulty understanding what to do or how to get started, work through the first problem with him or her.
- Review the work your child has done. This serves as reinforcement and provides further practice.
- Allow your child to use whatever writing instruments he or she prefers. For example, colored pencils can add variety and pleasure to drill work.
- Pay attention to the areas in which your child has the most difficulty. Provide extra guidance and exercises in those areas. Allowing children to use drawings and manipulatives, such as coins, tiles, game markers, or flash cards, can help them grasp difficult concepts more easily.
- Look for ways to make real-life applications to the skills being reinforced.

Practice 1

Look at each row of buttons. Decide what comes next and draw it in the space.

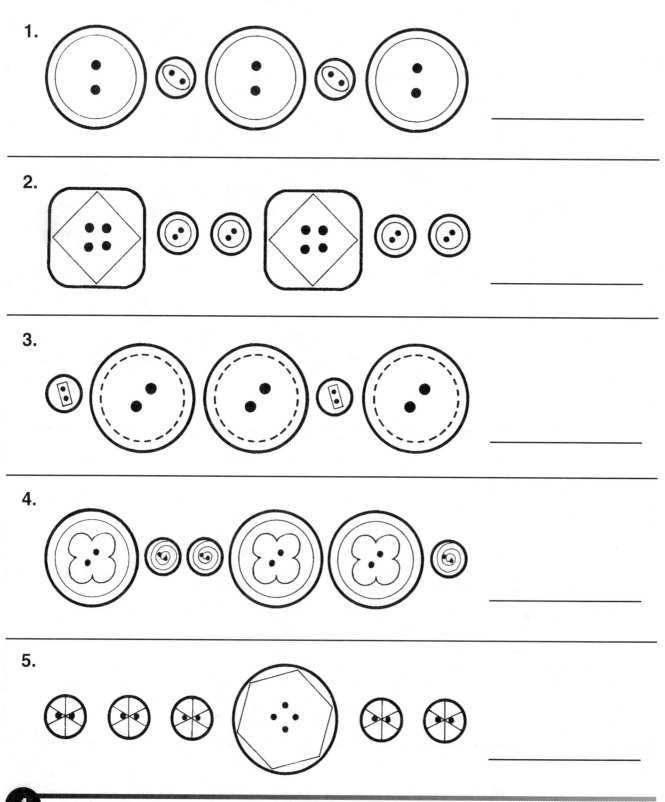

1.

2.

3.

4.

5.

Practice 2

Draw the next thing in each series.

1.

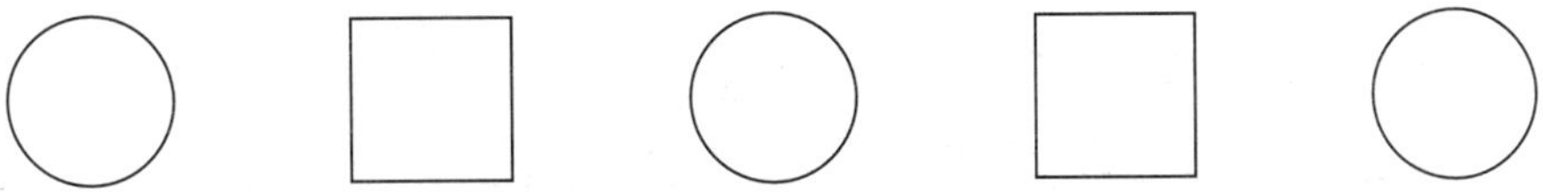

○ □ ○ □ ○

2.

♡ ☆ ♡ ☆ ♡

3.

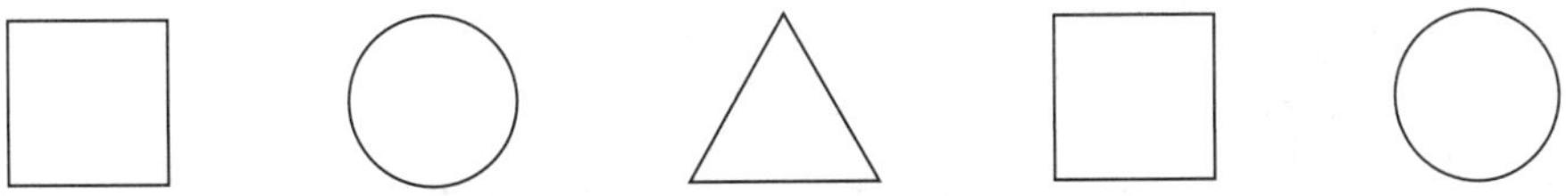

□ ○ △ □ ○

4.

☆ □ ☆ □ ☆

5.

○ ○ ▭ ○ ○

Practice 3

Draw the next thing in each series.

1. ○ □ □ ○ □ □ ○ □	
2. ♡ △ △ ♡ △ △ ♡ △	
3. ○ ☆ ○ ☆ ○ ☆ ○ ☆	
4. □ □ □ D D D ♡ ♡	
5. ○ △ ○ ▽ ○ △ ○ ▽	
6. ≡ ☆ □ ≡ ☆ □ ≡ ☆	
7. △ △ ☆ ♡ ♡ ☆ D D	
8. △ △ ○ ○ △ △ □ □ △	
9. ♡ ♡ ☆ ♡ ♡ □ ♡ ♡ ☆ ♡	
10. □ T □ ⊥ □ T □ ⊥	

Practice 4

Write the next letter in each series.

1. J K L M N O P _____

2. R S T U V W X Y _____

3. B A B B B C B D B E _____

4. A B B C C C D D D _____

5. X Y Z A B C X Y Z A _____

6. X X X Y Y Y Z Z _____

7. A E I O U A E I O U _____

8. M N O M N P M N Q _____

9. Z A A Z B B Z C _____

10. Z Y X W V U T S R Q _____

Practice 5

Directions: Write the next two numbers in each series.

1. 1, 2, 3, 4, 5, 6, 7, 8, ___ , ___

2. 2, 4, 6, 8, 10, 12, 14, 16, ___ , ___

3. 1, 3, 5, 7, 9, 11, 13, 15, ___ , ___

4. 1, 2, 3, 5, 6, 7, 9, 10, ___ , ___

5. 1, 1, 2, 2, 3, 3, 4, 4, ___ , ___

6. 9, 8, 7, 6, 5, 4, 3, 2, ___ , ___

7. 5, 10, 15, 20, 25, 30, 35, 40, ___ , ___

8. 1, 4, 7, 10, 13, 16, 19, 22, ___ , ___

9. 10, 20, 30, 40, 50, 60, 70, 80, ___ , ___

10. 18, 16, 14, 12, 10, 8, 6, 4, ___ , ___

11. 0, 1, 0, 2, 0, 3, 0, 4, ___ , ___

12. 5, 4, 3, 2, 1, 5, 4, 3, ___ , ___

13. 3, 1, 4, 5, 9, 14, 23, 37, ___ , ___

14. 3, 6, 9, 12, 9, 6, 3, 6, ___ , ___

#3320 Practice Makes Perfect: Graphs & Patterns

Practice 6

Write the next number in each series.

a. 1 2 3 4 5 6 7 8 _____

b. 2 4 6 8 10 12 14 16 _____

c. 5 10 15 20 25 30 _____

d. 2 3 4 6 7 8 10 11 _____

e. 2 2 3 3 4 4 5 5 _____

f. 1 3 1 4 1 5 1 6 1 _____

g. 8 7 6 5 4 3 2 _____

h. 3 2 6 4 9 6 12 8 _____

i. 1 5 9 13 17 21 25 _____

j. 10 20 30 40 50 60 70 80 _____

k. 20 18 16 14 12 10 8 _____

l. 0 1 0 3 0 5 0 7 _____

m. 6 5 4 3 2 1 6 5 4 _____

n. 3 6 9 12 18 21 24 _____

Practice 7

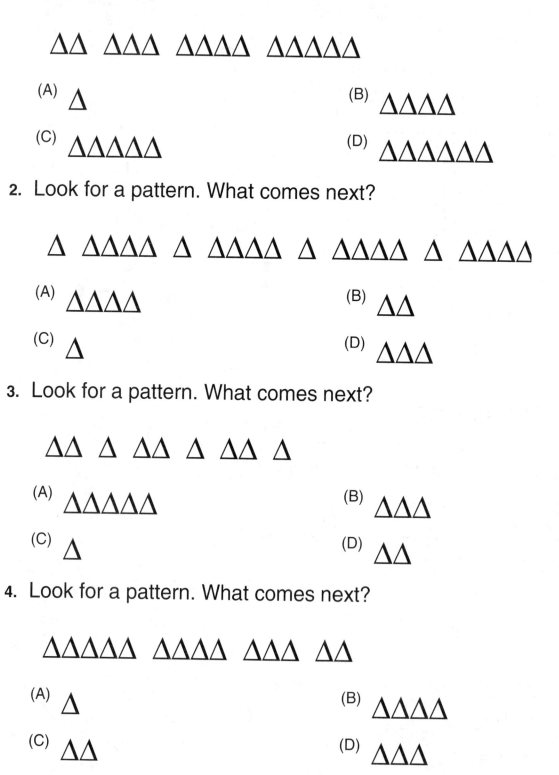

1. Look for a pattern. What comes next?

   △△ △△△ △△△△ △△△△△

   (A) △

   (B) △△△△

   (C) △△△△△

   (D) △△△△△△

2. Look for a pattern. What comes next?

   △ △△△△ △ △△△△ △ △△△△ △ △△△△

   (A) △△△△

   (B) △△

   (C) △

   (D) △△△

3. Look for a pattern. What comes next?

   △△ △ △△ △ △△ △

   (A) △△△△△

   (B) △△△

   (C) △

   (D) △△

4. Look for a pattern. What comes next?

   △△△△△ △△△△ △△△ △△

   (A) △

   (B) △△△△

   (C) △△

   (D) △△△

Practice 8

1. What comes next?

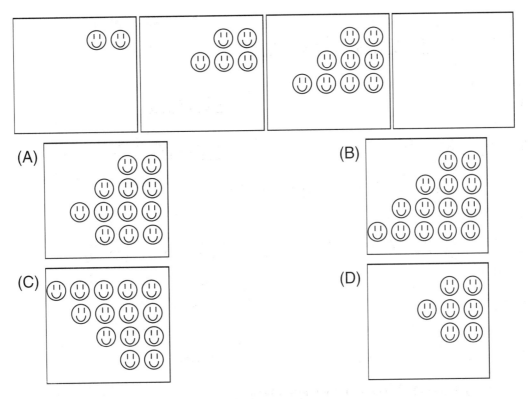

2. What comes next?

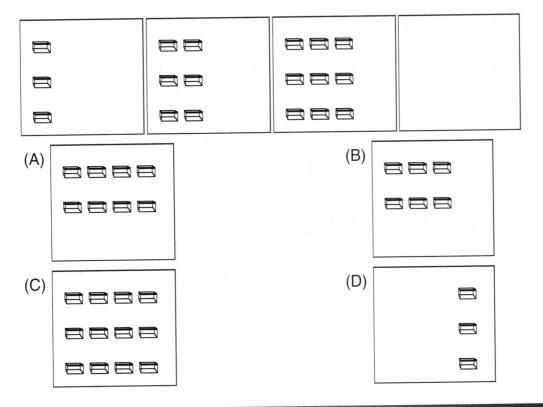

Practice 9

1. What comes next?

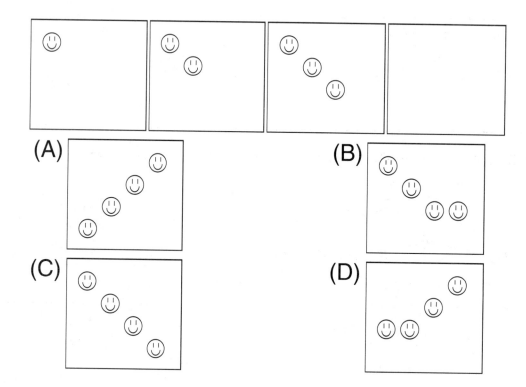

2. What comes next?

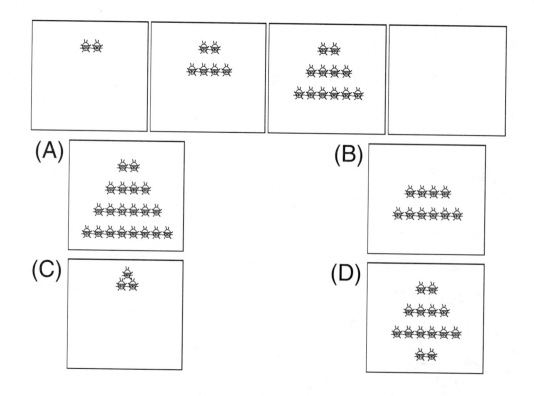

Practice 10 ౭ ☙ ౭ ☙ ౭ ☙ ౭ ☙ ౭ ☙ ౭ ☙ ౭ ౭ ☙

1. Look for a pattern. Find the next three numbers.

    6, 10, 14, ___, ___, ___

    (A) 17, 21, 25   (B) 15, 16, 17   (C) 19, 21, 27   (D) 18, 22, 26

2. Look for a pattern. Find the next three numbers.

    5, 9, 13, ___, ___, ___

    (A) 18, 20, 26   (B) 16, 20, 24   (C) 14, 15, 16   (D) 17, 21, 25

3. Look for a pattern. Find the next three numbers.

    26, 23, 20, ___, ___, ___

    (A) 18, 13, 12   (B) 19, 18, 17   (C) 17, 14, 11   (D) 16, 13, 10

4. Look for a pattern. Find the next three numbers.

    2, 6, 10, ___, ___, ___

    (A) 11, 12, 13   (B) 15, 17, 23   (C) 13, 17, 21   (D) 14, 18, 22

5. Look for a pattern. Find the next three numbers.

    4, 9, 14, ___, ___, ___

    (A) 18, 23, 28   (B) 19, 24, 29   (C) 15, 16, 17   (D) 20, 23, 30

Practice 11

1. Look for a pattern. Find the next three numbers.

   17, 15, 13, ___, ___, ___

   (A) 11, 9, 7   (B) 10, 8, 6

   (C) 12, 8, 8   (D) 12, 11, 10

4. Look for a pattern. Find the next three numbers.

   21, 18, 15, ___, ___, ___

   (A) 13, 8, 7   (B) 11, 8, 5

   (C) 14, 13, 12   (D) 12, 9, 6

2. Look for a pattern. Find the next three numbers.

   7, 13, 19, ___, ___, ___

   (A) 26, 30, 38

   (B) 20, 21, 22

   (C) 24, 30, 36

   (D) 25, 31, 37

5. Look for a pattern. Find the next three numbers.

   32, 28, 24, ___, ___, ___

   (A) 19, 15, 11

   (B) 20, 16, 12

   (C) 21, 15, 13

   (D) 23, 22, 21

3. Look for a pattern. Find the next three numbers.

   32, 27, 22, ___, ___, ___

   (A) 17, 12, 7

   (B) 18, 11, 8

   (C) 16, 11, 6

   (D) 21, 20, 19

6. Look for a pattern. Find the next three numbers.

   16, 14, 12, ___, ___, ___

   (A) 11, 10, 9   (B) 11, 7, 7

   (C) 9, 7, 5   (D) 10, 8, 6

Practice 12

Write the correct numeral in each blank.

1. 16, 17, 18, _____

2. 12, 13, _____

3. 4, 5, _____

4. 31, 32, _____

5. 40, 41, _____

6. 60, _____, 62

7. 42, _____, 44

8. 49, _____, 51

9. 56, _____, 58

10. _____, 82, 83

11. 77, 78, _____

12. 90, _____, 92

13. _____, 36, 37

14. _____, 52, 53

15. 97, 98, _____

16. 70, _____, 72

17. 39, _____, 41

18. 92, _____, 94

19. 85, _____, 87

20. 53, _____, 55

In the space below, make up some of your own counting puzzles.

Practice 13

1. One box is empty. Which drawing completes the pattern?

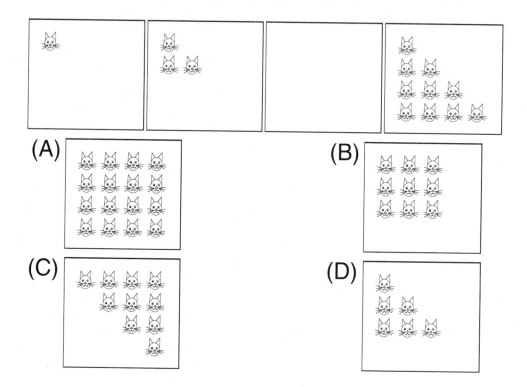

2. One box is empty. Which drawing completes the pattern?

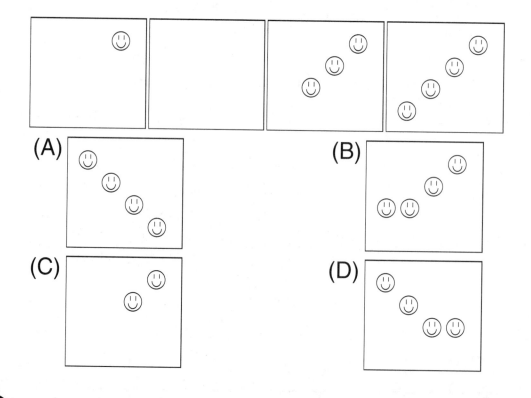

Practice 14

1. Find the missing shape.

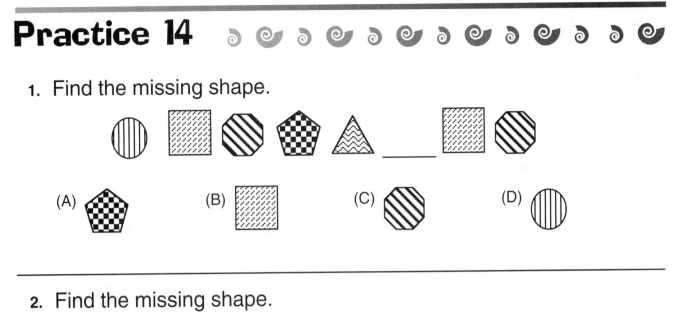

2. Find the missing shape.

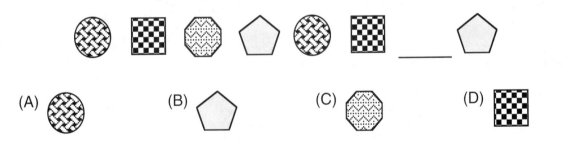

3. Find the missing shape.

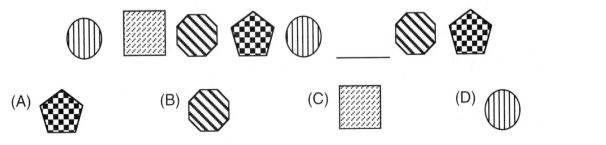

4. Find the missing shape.

Practice 15

1. Find the missing shape.

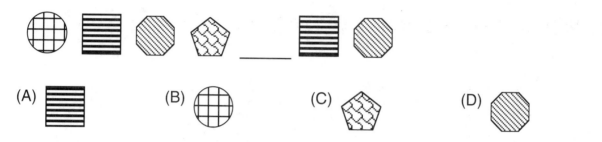

2. Find the missing shape.

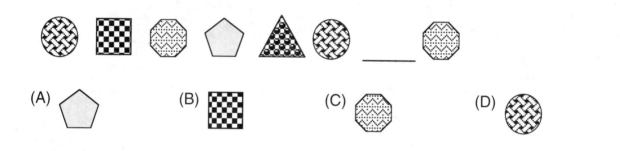

3. Find the missing shape.

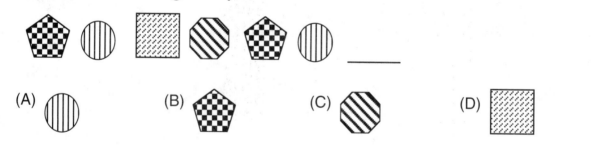

4. Find the missing shape.

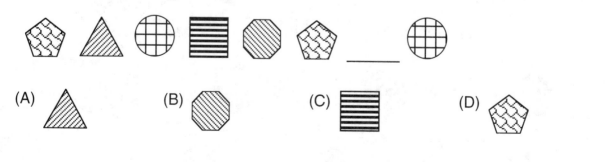

Practice 16

1. Which shape belongs with the group?

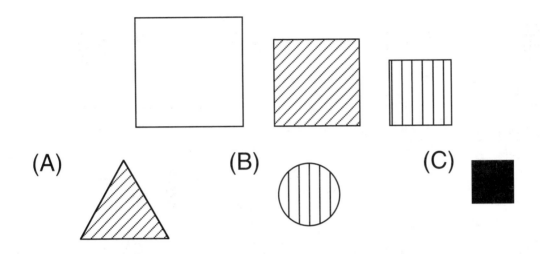

(A)   (B)   (C)

2. Which shape belongs with the group?

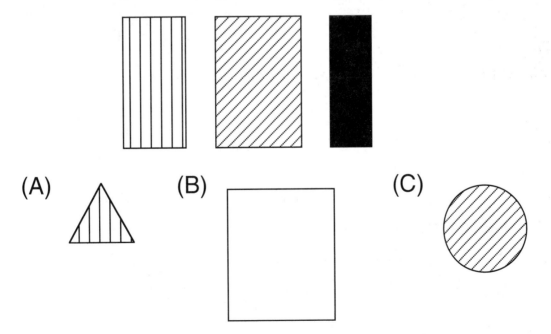

(A)   (B)   (C)

Practice 17

1. Look for a pattern. Find the missing number.

   9, 12, ____ , 18

   (A) 16   (B) 15

   (C) 14   (D) 13

2. Look for a pattern. Find the missing number.

   7, 13, ____ , 25

   (A) 19   (B) 18

   (C) 14   (D) 20

3. Look for a pattern. Find the missing number.

   3, 11, ____ , 27

   (A) 12   (B) 20

   (C) 18   (D) 19

4. Look for a pattern. Find the missing number.

   4, 8, ____ , 16

   (A) 11   (B) 9   (C) 12   (D) 13

5. Look for a pattern. Find the missing number.

   8, 13, ____ , 23

   (A) 19   (B) 18

   (C) 14   (D) 17

6. Look for a pattern. Find the missing number.

   5, 14, ____ , 32

   (A) 23   (B) 22

   (C) 24   (D) 15

7. Look for a pattern. Find the missing number.

   2, 9, ____ , 23

   (A) 10   (B) 17

   (C) 15   (D) 16

8. Look for a pattern. Find the missing number.

   6, 9, ____ , 15

   (A) 13   (B) 10

   (C) 11   (D) 12

Practice 18

1. What number goes in the circle?

+	1	2	3	4	5	6	7	8	9
1	2	3	4	5	6	7	8	9	10
2	3	4	5	6	7	8	9	10	11
3	4	5	6	7	8	9	10	11	12
4	5	6	7	8	9	10	11	12	13
5	6	7	8	9	10	11	12	13	14
6	7	8	9	○	11	12	13	14	15
7	8	9	10	11	12	13	14	15	16
8	9	10	11	12	13	14	15	16	17
9	10	11	12	13	14	15	16	17	18

(A) 4 (B) 10 (C) 11 (D) 6

2. What number goes in the circle?

+	1	2	3	4	5	6	7	8	9
1	2	3	4	5	6	7	8	9	10
2	3	4	5	6	7	8	9	10	11
3	4	5	6	7	8	9	10	11	12
4	5	6	7	8	9	10	11	12	13
5	6	7	8	9	10	11	12	13	14
6	7	8	9	10	11	12	13	14	15
7	8	9	10	11	12	13	14	15	16
8	9	10	○	12	13	14	15	16	17
9	10	11	12	13	14	15	16	17	18

(A) 11 (B) 3 (C) 8 (D) 10

Practice 19

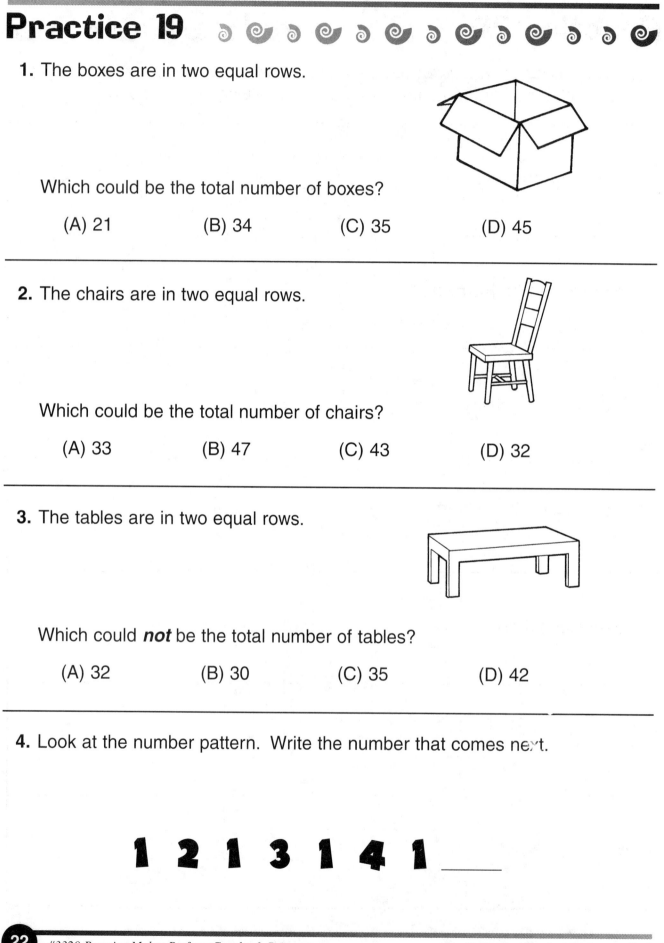

1. The boxes are in two equal rows.

Which could be the total number of boxes?

(A) 21   (B) 34   (C) 35   (D) 45

2. The chairs are in two equal rows.

Which could be the total number of chairs?

(A) 33   (B) 47   (C) 43   (D) 32

3. The tables are in two equal rows.

Which could *not* be the total number of tables?

(A) 32   (B) 30   (C) 35   (D) 42

4. Look at the number pattern. Write the number that comes next.

1 2 1 3 1 4 1 ____

Practice 20

Make your own patterns using the following objects. Use any number of objects. You can make the patterns easy or hard. The challenge is up to you.

different shapes
letters of the alphabet
different flowers
numbers
favorite fruits
animals
favorite toys

Practice 21

1. Use the tally chart to solve.

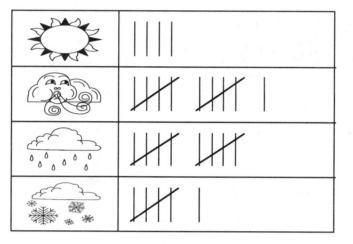

Weather for December

How many more snowy days were there than sunny days?

(A) 8   (B) 5   (C) 2   (D) 3

2. Use the tally chart to solve.

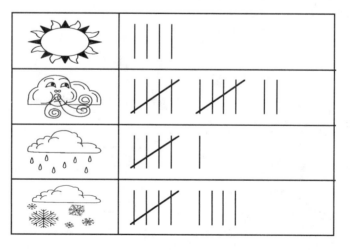

Weather for December

How many fewer rainy days were there than snowy days?

(A) 6   (B) 9   (C) 3   (D) 15

Practice 22 ୬ ௰ ୬ ௰ ୬ ௰ ୬ ௰ ୬ ௰ ୬ ୬ ௰

The students in Mr. Lockwood's class took a class vote to find out what subject was liked most by class members. Mr. Lockwood kept a record of the votes for each subject on a tally chart.

Read the chart. Then answer the questions.

Our Favorite Subjects	
Subject	**Tally of Votes**
mathematics	|||
art	||||
history		
science	~~||||~~
music	||
reading	~~||||~~	
writing	|||
physical education	~~||||~~
health		

1. What subject is liked the most? _____

2. How many people liked it the most? _____

3. What subjects tied for second place? _____

4. Which subjects tied with writing? _____

5. How many liked mathematics more than history? _____

Practice 23

One type of graph that gives us information is called a pictograph. In a pictograph, pictures are used instead of numbers.

1. How many students chose birds as their favorite animal?

Favorite Animals

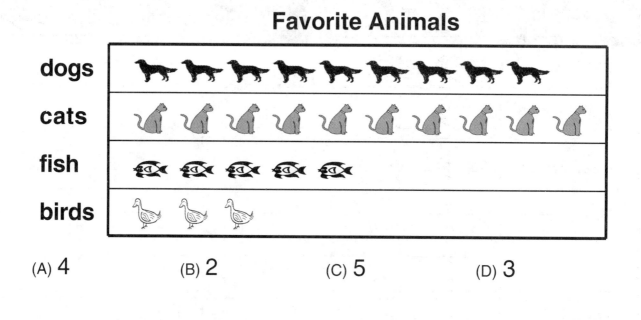

(A) 4   (B) 2   (C) 5   (D) 3

2. How many students chose fish as their favorite animal?

Favorite Animals

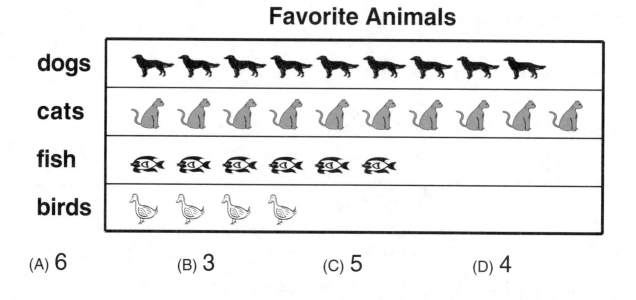

(A) 6   (B) 3   (C) 5   (D) 4

Practice 24

Here is a pictograph that shows the number of fish caught each day at Canyon Lake.

Daily Fish Catch at Canyon Lake	
Sunday	🐟 🐟 🐟 🐟 🐟 🐟 🐟
Monday	🐟 🐟
Tuesday	🐟
Wednesday	🐟 🐟 🐟
Thursday	🐟 🐟
Friday	🐟 🐟 🐟 🐟 🐟
Saturday	🐟 🐟 🐟 🐟 🐟 🐟
KEY: 🐟 = 10 fish	

1. On what day were the most fish caught? _____

2. How many fish were caught on this day? _____

3. On what day were 50 fish caught? _____

4. On what day were the fewest fish caught? _____

5. Were the same number of fish caught on Monday and Thursday? _____

6. How many fish were caught on both Saturday and Sunday? _____

Practice 25

Look at the picture below. Find the faces and plot them on the graph. Then, answer the questions.

How many?

Circle the correct picture.

MORE

FEWER

FEWER

MORE

Practice 26

Ask your family or classmates whether they like cats, dogs, or birds best. Fill in the graph to show the animal that each person chose.

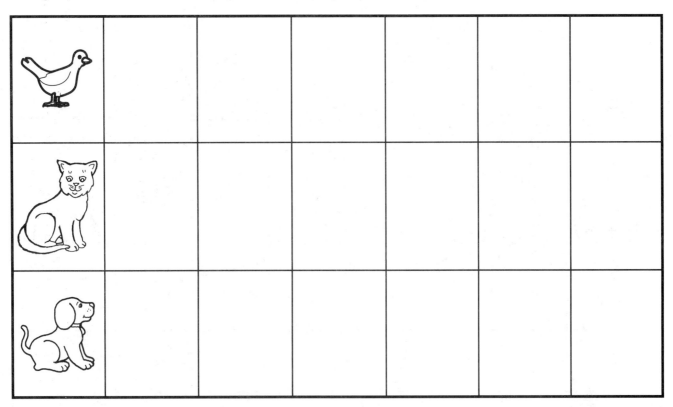

1. How many people liked birds best? _____

2. How many people liked cats best? _____

3. How many people liked dogs best? _____

4. Which animal did most people like the best? _____

5. Which animal did the fewest number of people like? _____

Write a sentence about the results.

_____

- -

_____

- -

_____

Practice 27 ა ❧ ა ❧ ა ❧ ა ❧ ა ❧ ა ❧ ა ❧ ა ❧

Look at these pictures of Joe, Spud, Helena, and Joanna. Count them to complete the graph. Fill in the spaces on the graph.

		1	**2**	**3**	**4**	**5**	**6**	**7**
a.	Joe							
b.	Joanna							
c.	Spud							
d.	Helena							

Practice 28

The graph tells how many of each animal there are. Use the graph to complete the addition problems below.

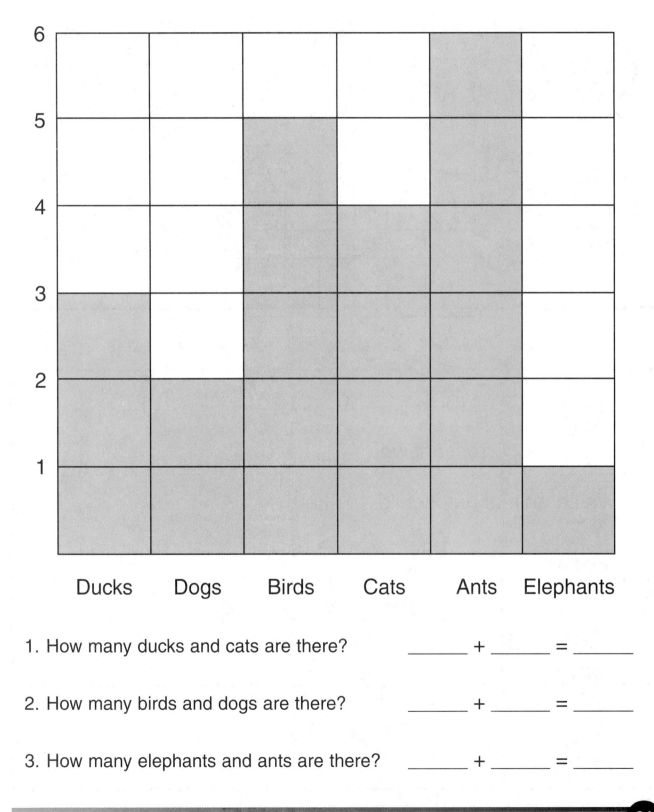

1. How many ducks and cats are there?   _____ + _____ = _____

2. How many birds and dogs are there?   _____ + _____ = _____

3. How many elephants and ants are there?   _____ + _____ = _____

Practice 29

1. How many children like football the best?

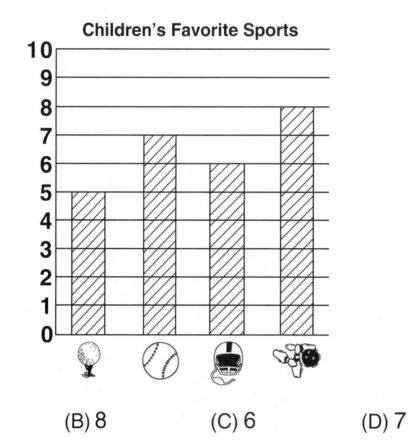

(A) 5   (B) 8   (C) 6   (D) 7

2. Circle the correct answer.

Which animal did 2 children like?

horse   cat   bird   dog

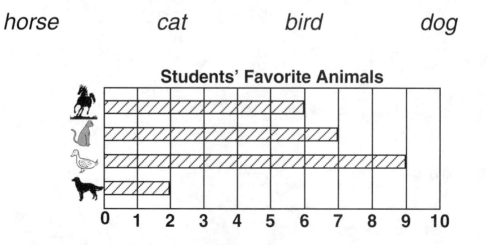

Practice 30

1. How many children like the 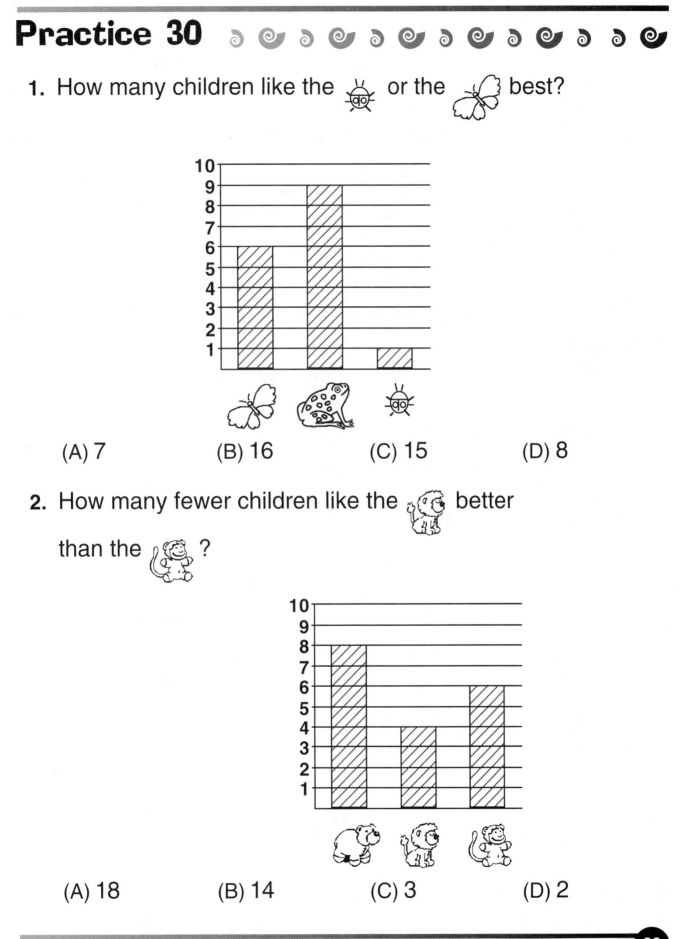 or the ☺ best?

(A) 7   (B) 16   (C) 15   (D) 8

2. How many fewer children like the 🦁 better

than the 🐵 ?

(A) 18   (B) 14   (C) 3   (D) 2

Practice 31

1. Use the graph to answer the question.

How many quarters does Bart have?

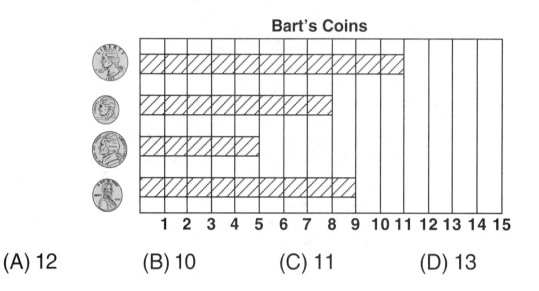

Bart's Coins

(A) 12   (B) 10   (C) 11   (D) 13

2. Use the graph to answer the question.

Which is worth the most?

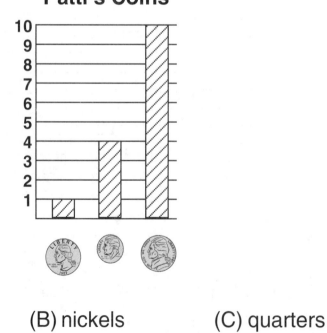

Patti's Coins

(A) dimes   (B) nickels   (C) quarters

Practice 32

1. Use the bar graph to answer the question.
   How many fewer pens are there than boxes of crayons?_____

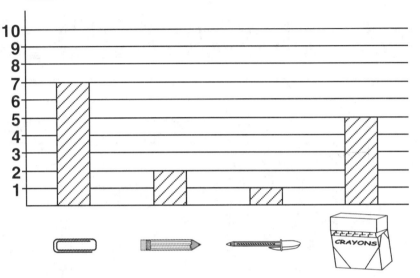

School Supplies in Mr. Saunder's Room

2. Use the bar graph to answer the question.
   How many more paper clips are there than pencils?_____

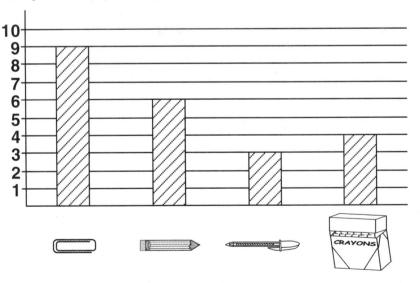

School Supplies in Mr. Saunder's Room

Practice 33

The classes at Barnsdale Elementary School kept a bar graph of the number of books each grade read for a week.

Study this bar graph of their reading and answer the questions below.

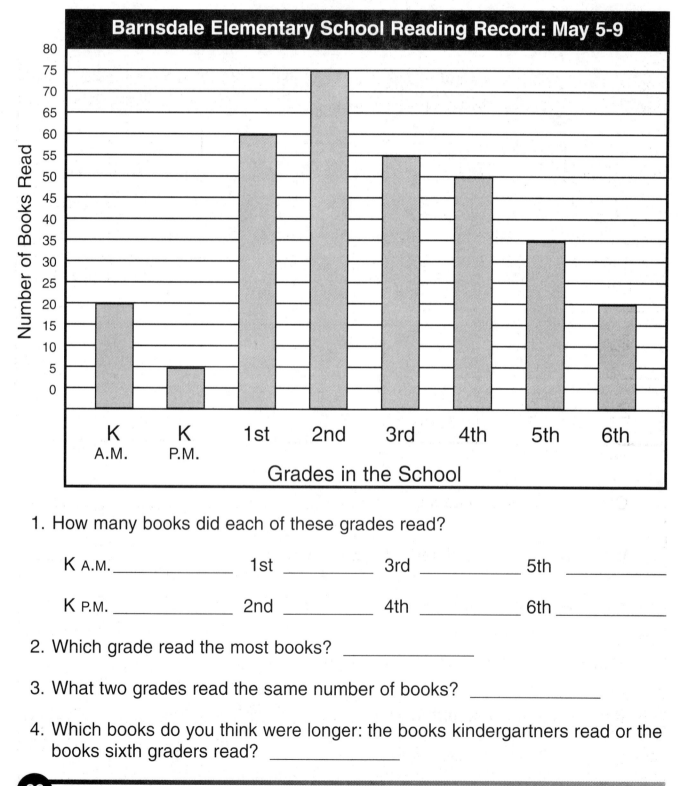

1. How many books did each of these grades read?

K A.M. _____   1st _____   3rd _____   5th _____

K P.M. _____   2nd _____   4th _____   6th _____

2. Which grade read the most books? _____

3. What two grades read the same number of books? _____

4. Which books do you think were longer: the books kindergartners read or the books sixth graders read? _____

Practice 34 ⟑ ⟑ ⟑ ⟑ ⟑ ⟑ ⟑ ⟑ ⟑ ⟑ ⟑ ⟑ ⟑

A calendar helps us get organized. A calendar will show us what month it is and what the date is. Use this calendar to answer the following questions.

August

Sunday	Monday	Tuesday	Wednesday	Thursday	Friday	Saturday
1	2	3 Christopher	4	5	6	7 Karen
8	9	10	11	12 Alyssa	13	14
15 Billy	16	17	18	19	20	21
22	23	24	25 Alexander	26	27	28
29	30	31 Feather				

1. On what day is Christopher's birthday? _____

2. What date (month and day) is Alyssa's birthday? _____

3. Alexander's birthday is on August 25. What day is that? _____

4. How many birthdays are on Tuesdays? _____

5. Which days have no birthdays at all? _____

Practice 35

A Venn diagram is a type of diagram that uses circles to show how things are related to each other. The overlapping parts of the circles show what things the circles have in common.

Look at this diagram of the activities of two kindergarten classes that share the same room.

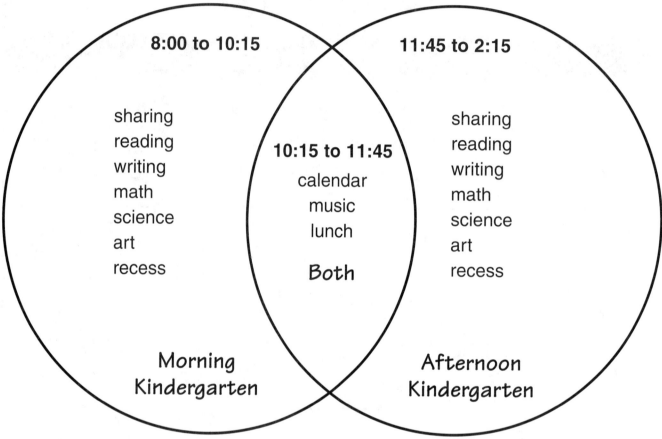

8:00 to 10:15

11:45 to 2:15

sharing
reading
writing
math
science
art
recess

10:15 to 11:45

calendar
music
lunch

Both

sharing
reading
writing
math
science
art
recess

Morning
Kindergarten

Afternoon
Kindergarten

1. By reading this diagram, can you tell what time of day the two classes are together in the same room? _____ What time? _____

2. What activities do both classes do together? _____

_____

3. What activities do the classes each do by themselves?

_____

_____

_____

Practice 36

One type of graph that gives us information is called a circle graph. In a circle graph, you can show how things are divided into the parts of a whole.

Read this circle graph about where Derek spends the hours in one day.

My 24-Hour Day

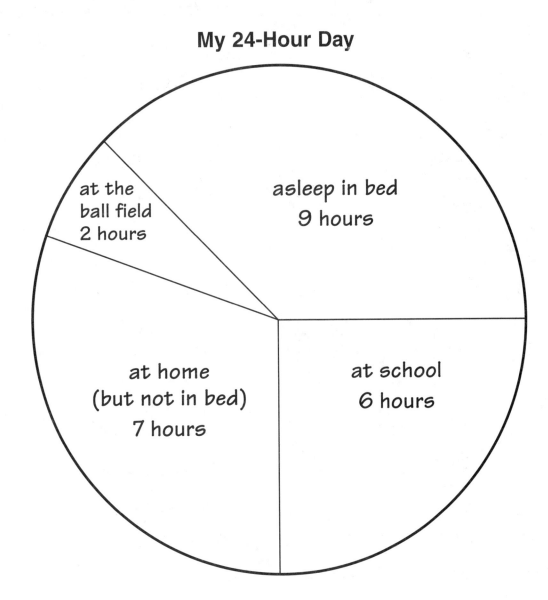

1. Use a red crayon to color the place Derek spends six hours a day.

2. Use a blue crayon to color Derek's sleeping time.

3. Use a green crayon to color the time Derek spends on the ball field.

4. Use a yellow crayon to color the time Derek is **not** sleeping while he is at home.

Test Practice 1

1. Look for a pattern. What comes next?

△ △△△△ △ △△△△ △ △△△△ △ △△△△

(A) △△△△

(B) △

(C) △△△

(D) △△

2. Find the missing shape.

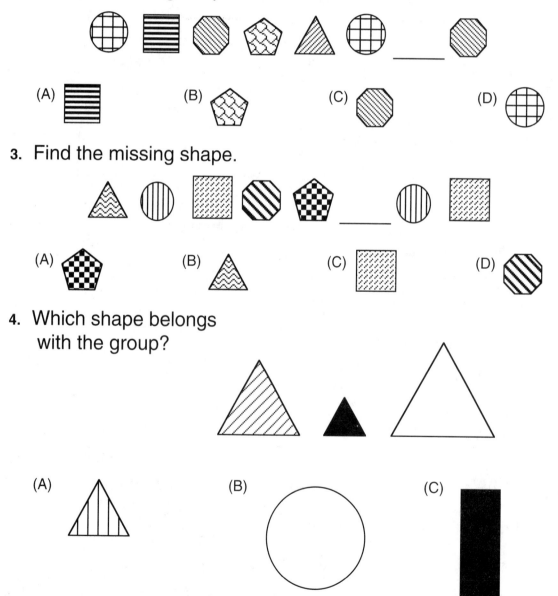

(A) ▤

(B) ⬠

(C) ⯃

(D) ⊕

3. Find the missing shape.

(A) ⬠

(B) △

(C) ▢

(D) ⬡

4. Which shape belongs with the group?

(A)

(B)

(C)

Test Practice 2

1. How many students chose dogs as their favorite animal?

Favorite Animals

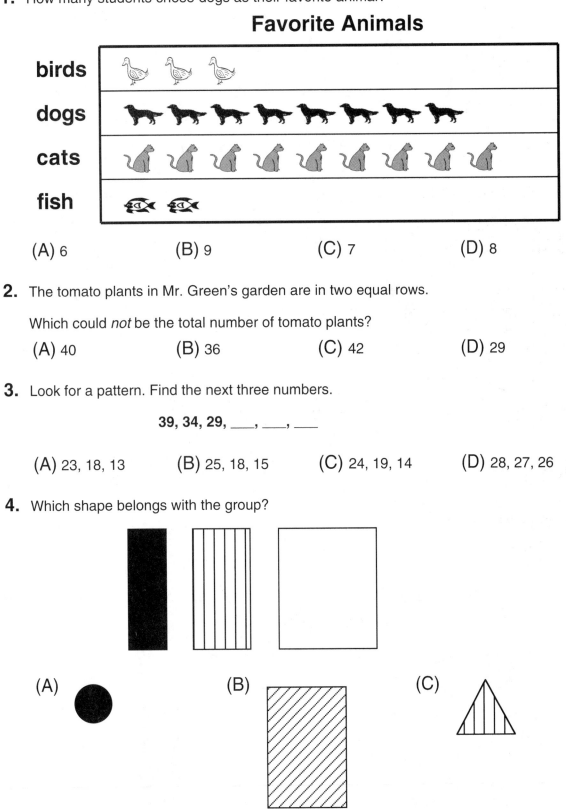

(A) 6   (B) 9   (C) 7   (D) 8

2. The tomato plants in Mr. Green's garden are in two equal rows.

Which could *not* be the total number of tomato plants?

(A) 40   (B) 36   (C) 42   (D) 29

3. Look for a pattern. Find the next three numbers.

39, 34, 29, ___, ___, ___

(A) 23, 18, 13   (B) 25, 18, 15   (C) 24, 19, 14   (D) 28, 27, 26

4. Which shape belongs with the group?

(A)   (B)   (C)

Test Practice 3

1. Look for a pattern. Find the next three numbers.

    9, 11, 13, ___, ___, ___

    (A) 15, 17, 19   (B) 16, 16, 20   (C) 14, 16, 18   (D) 14, 15, 16

2. How many fewer children like the 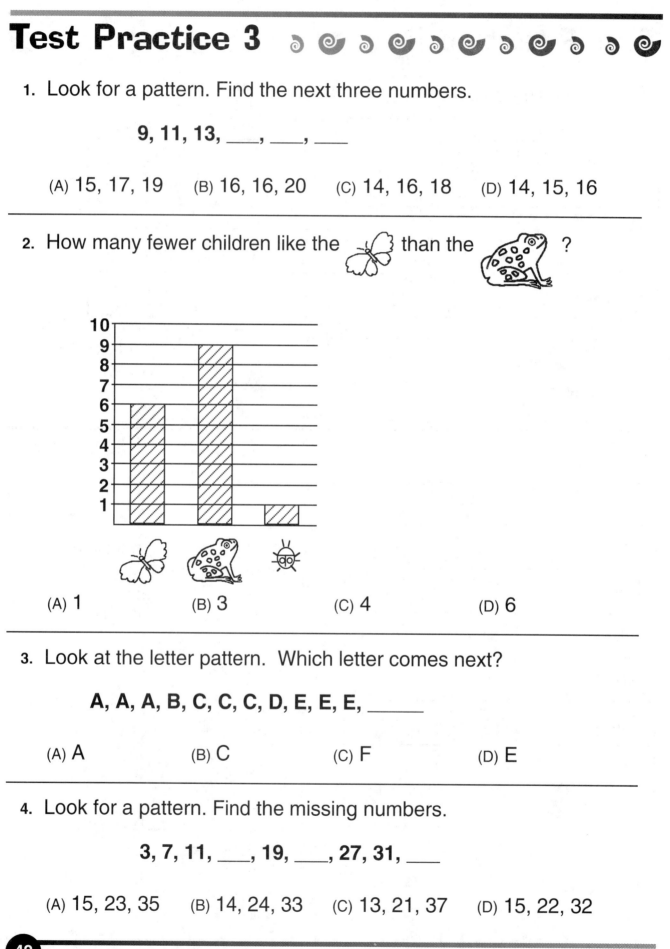 than the ?

    (A) 1   (B) 3   (C) 4   (D) 6

3. Look at the letter pattern. Which letter comes next?

    A, A, A, B, C, C, C, D, E, E, E, _____

    (A) A   (B) C   (C) F   (D) E

4. Look for a pattern. Find the missing numbers.

    3, 7, 11, ___, 19, ___, 27, 31, ___

    (A) 15, 23, 35   (B) 14, 24, 33   (C) 13, 21, 37   (D) 15, 22, 32

Test Practice 4 ⟋ ⟍ ⟋ ⟍ ⟋ ⟍ ⟋ ⟍ ⟋ ⟍ ⟋ ⟍

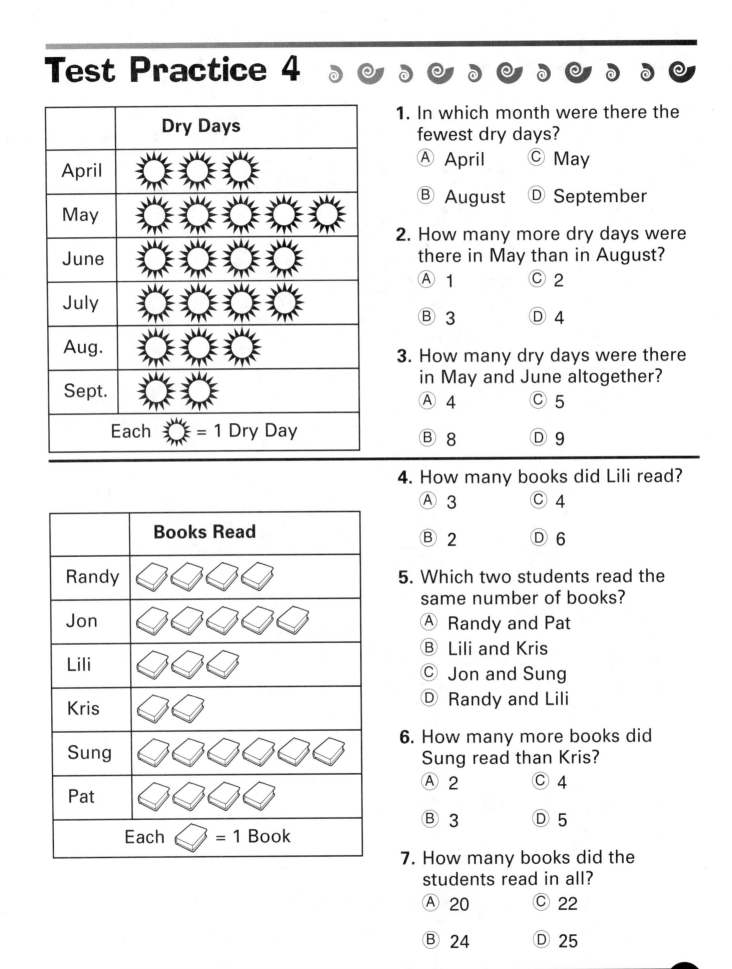

	Dry Days
April	☀ ☀ ☀
May	☀ ☀ ☀ ☀ ☀
June	☀ ☀ ☀ ☀
July	☀ ☀ ☀ ☀
Aug.	☀ ☀ ☀
Sept.	☀ ☀

Each ☀ = 1 Dry Day

	Books Read
Randy	📖 📖 📖 📖
Jon	📖 📖 📖 📖 📖
Lili	📖 📖 📖
Kris	📖 📖
Sung	📖 📖 📖 📖 📖 📖
Pat	📖 📖 📖 📖

Each 📖 = 1 Book

1. In which month were there the fewest dry days?
   - Ⓐ April
   - Ⓒ May
   - Ⓑ August
   - Ⓓ September

2. How many more dry days were there in May than in August?
   - Ⓐ 1
   - Ⓒ 2
   - Ⓑ 3
   - Ⓓ 4

3. How many dry days were there in May and June altogether?
   - Ⓐ 4
   - Ⓒ 5
   - Ⓑ 8
   - Ⓓ 9

4. How many books did Lili read?
   - Ⓐ 3
   - Ⓒ 4
   - Ⓑ 2
   - Ⓓ 6

5. Which two students read the same number of books?
   - Ⓐ Randy and Pat
   - Ⓑ Lili and Kris
   - Ⓒ Jon and Sung
   - Ⓓ Randy and Lili

6. How many more books did Sung read than Kris?
   - Ⓐ 2
   - Ⓒ 4
   - Ⓑ 3
   - Ⓓ 5

7. How many books did the students read in all?
   - Ⓐ 20
   - Ⓒ 22
   - Ⓑ 24
   - Ⓓ 25

Test Practice 5

Sue made a graph to show the animals she saw in her back yard. Use the graph to answer the questions. If the answer is not given, choose "Not Here."

1. How many dogs did Sue see?

   2   3   6   Not Here
   (A)   (B)   (C)   (D)

2. How many turtles did Sue see?

   0   3   5   Not Here
   (A)   (B)   (C)   (D)

3. How many animals did Sue see in all?

   7   16   10   Not Here
   (A)   (B)   (C)   (D)

The graph shows the number of balloons lost at a carnival each day. Use it to answer the questions below. If the answer is not given, choose "Not Here."

Mon.	O O O O
Tues.	O O O O O O O
Wed.	O

4. How many balloons were lost on Monday?

   4   5   6   Not Here
   (A)   (B)   (C)   (D)

5. How many balloons were lost on Tuesday?

   2   4   6   Not Here
   (A)   (B)   (C)   (D)

6. How many balloons were lost on Tuesday and Wednesday?

   7   8   9   Not Here
   (A)   (B)   (C)   (D)

Test Practice 6

Choose the next shape or number in the pattern. Choose "Not Here" if the number or pattern is not one of the choices given.

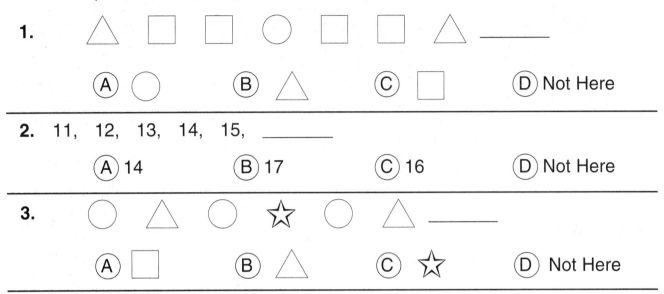

1. △ □ □ ○ □ □ △ _____

    (A) ○   (B) △   (C) □   (D) Not Here

2. 11, 12, 13, 14, 15, _____

    (A) 14   (B) 17   (C) 16   (D) Not Here

3. ○ △ ○ ☆ ○ △ _____

    (A) □   (B) △   (C) ☆   (D) Not Here

Mr. Todd's class voted on their favorite ice-cream flavors. The graph shows how many students chose each flavor. Use the graph to answer the questions.

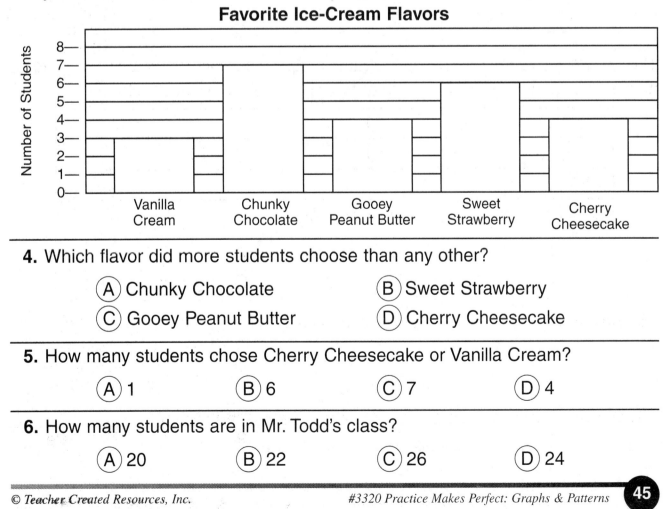

Favorite Ice-Cream Flavors

4. Which flavor did more students choose than any other?

    (A) Chunky Chocolate   (B) Sweet Strawberry

    (C) Gooey Peanut Butter   (D) Cherry Cheesecake

5. How many students chose Cherry Cheesecake or Vanilla Cream?

    (A) 1   (B) 6   (C) 7   (D) 4

6. How many students are in Mr. Todd's class?

    (A) 20   (B) 22   (C) 26   (D) 24

Answer Sheet

Test Practice 1	Test Practice 2	Test Practice 3
1. (A) (B) (C) (D)	1. (A) (B) (C) (D)	1. (A) (B) (C) (D)
2. (A) (B) (C) (D)	2. (A) (B) (C) (D)	2. (A) (B) (C) (D)
3. (A) (B) (C) (D)	3. (A) (B) (C) (D)	3. (A) (B) (C) (D)
4. (A) (B) (C) (D)	4. (A) (B) (C) (D)	4. (A) (B) (C) (D)

Test Practice 4	Test Practice 5	Test Practice 6
1. (A) (B) (C) (D)	1. (A) (B) (C) (D)	1. (A) (B) (C) (D)
2. (A) (B) (C) (D)	2. (A) (B) (C) (D)	2. (A) (B) (C) (D)
3. (A) (B) (C) (D)	3. (A) (B) (C) (D)	3. (A) (B) (C) (D)
4. (A) (B) (C) (D)	4. (A) (B) (C) (D)	4. (A) (B) (C) (D)
5. (A) (B) (C) (D)	5. (A) (B) (C) (D)	5. (A) (B) (C) (D)
6. (A) (B) (C) (D)	6. (A) (B) (C) (D)	6. (A) (B) (C) (D)
7. (A) (B) (C) (D)		

Answer Key

Test Practice 1	Test Practice 2	Test Practice 3
1. Ⓐ ● Ⓒ Ⓓ	1. Ⓐ Ⓑ Ⓒ ●	1. ● Ⓑ Ⓒ Ⓓ
2. ● Ⓑ Ⓒ Ⓓ	2. Ⓐ Ⓑ Ⓒ ●	2. Ⓐ ● Ⓒ Ⓓ
3. Ⓐ ● Ⓒ Ⓓ	3. Ⓐ Ⓑ ● Ⓓ	3. Ⓐ Ⓑ ● Ⓓ
4. ● Ⓑ Ⓒ Ⓓ	4. Ⓐ ● Ⓒ Ⓓ	4. ● Ⓑ Ⓒ Ⓓ

Test Practice 4	Test Practice 5	Test Practice 6
1. Ⓐ Ⓑ Ⓒ ●	1. Ⓐ Ⓑ ● Ⓓ	1. Ⓐ Ⓑ ● Ⓓ
2. Ⓐ Ⓑ ● Ⓓ	2. Ⓐ ● Ⓒ Ⓓ	2. Ⓐ Ⓑ ● Ⓓ
3. Ⓐ Ⓑ Ⓒ ●	3. Ⓐ ● Ⓒ Ⓓ	3. Ⓐ Ⓑ Ⓒ ●
4. ● Ⓑ Ⓒ Ⓓ	4. ● Ⓑ Ⓒ Ⓓ	4. ● Ⓑ Ⓒ Ⓓ
5. ● Ⓑ Ⓒ Ⓓ	5. Ⓐ Ⓑ Ⓒ ●	5. Ⓐ Ⓑ ● Ⓓ
6. Ⓐ Ⓑ ● Ⓓ	6. Ⓐ ● Ⓒ Ⓓ	6. Ⓐ Ⓑ Ⓒ ●
7. Ⓐ ● Ⓒ Ⓓ		

Answer Key ෨ ෨ ෨ ෨ ෨ ෨ ෨ ෨ ෨ ෨ ෨ ෨ ෨

Page 4
1. small button
2. large button
3. large button
4. small button
5. small button

Page 5
1. square
2. star
3. triangle
4. square
5. rectangle

Page 6
1. square
2. triangle
3. circle
4. heart
5. circle
6. square
7. star
8. triangle
9. heart
10. square

Page 7
1. Q
2. Z
3. B
4. E
5. B
6. Z
7. A
8. M
9. C
10. P

Page 8
1. 9, 10
2. 18, 20
3. 17, 19
4. 11, 13
5. 5, 5
6. 1, 0
7. 45, 50
8. 25, 28
9. 90, 100
10. 2, 0
11. 0, 5
12. 2, 1
13. 60, 97
14. 9, 12

Page 9
a. 9
b. 18
c. 35
d. 12
e. 6
f. 7
g. 1
h. 15
i. 29
j. 90
k. 6
l. 0
m. 3
n. 27

Page 10
1. D
2. C
3. D
4. A

Page 11
1. B
2. C

Page 12
1. C
2. A

Page 13
1. D
2. D
3. C
4. D
5. B

Page 14
1. A
2. D
3. A
4. D
5. B
6. D

Page 15
1. 19
2. 14
3. 6
4. 33
5. 42
6. 61
7. 43
8. 50
9. 57
10. 81
11. 79
12. 91
13. 35
14. 51
15. 99
16. 71
17. 40
18. 93
19. 86
20. 54

Page 16
1. D
2. C

Page 17
1. D
2. C
3. C
4. C

Page 18
1. B
2. B
3. D
4. A

Page 19
1. C
2. B

Page 20
1. B
2. A
3. D
4. C
5. B
6. A
7. D
8. D

Page 21
1. B
2. A

Page 22
1. B
2. D
3. C
4. 5

Page 24
1. C
2. C

Page 25
1. reading
2. 6
3. science, physical education
4. mathematics
5. 2

Page 26
1. D
2. A

Page 27
1. Sunday
2. 70
3. Friday
4. Tuesday
5. yes
6. 130

Page 28

Page 29
Graphs will vary.

Page 30

Page 31
1. 7
2. 7
3. 7

Page 32
1. C
2. dog

Page 33
1. A
2. D

Page 34
1. C
2. B

Page 35
1. 4
2. 3

Page 36
1. K (A.M.) 20
   K (P.M.) 5
   1st 60
   2nd 75
   3rd 55
   4th 50
   5th 35
   6th 20
2. 2nd
3. K (A.M.) and 6th
4. 6th graders

Page 37
1. Tuesday
2. August 12
3. Wednesday
4. 2
5. Monday, Friday

Page 38
1. yes, 10:15 to 11:45
2. calendar, music, lunch
3. sharing, reading, writing, math, science, art, recess

Page 40
1. B
2. A
3. B
4. A

Page 41
1. D
2. D
3. C
4. B

Page 42
1. A
2. B
3. C
4. A

Page 43
1. D
2. C
3. D
4. A
5. A
6. C
7. B

Page 44
1. C
2. B
3. B
4. A
5. D
6. B

Page 45
1. C
2. C
3. D
4. A
5. C
6. D

#3320 Practice Makes Perfect: Graphs & Patterns